Light and Heat

Copyright © by Harcourt, Inc.

All rights reserved. No part of this publication may be reproduced or transmitted in any form or by any means, electronic or mechanical, including photocopy, recording, or any information storage and retrieval system, without permission in writing from the publisher.

Requests for permission to make copies of any part of the work should be addressed to School Permissions and Copyrights, Harcourt, Inc., 6277 Sea Harbor Drive, Orlando, Florida 32887-6777. Fax: 407-345-2418.

HARCOURT and the Harcourt Logo are trademarks of Harcourt, Inc., registered in the United States of America and/or other jurisdictions.

Printed in México

ISBN-13: 978-0-15-362018-8
ISBN-10: 0-15-362018-8

2 3 4 5 6 7 8 9 10 805 16 15 14 13 12 11 10 09 08

SCHOOL PUBLISHERS

Visit *The Learning Site!*
www.harcourtschool.com

Forms of Energy

Energy can cause matter to move or change.
Light is a form of energy.
Heat is also a form of energy.
Sound is a form of energy.

Where Energy Comes From

moving water

Solar energy is energy from the sun.
Some energy comes from wind.
Energy also comes from moving water.
Energy comes from fuels like coal and oil.

Light

Light is a form of energy that lets you see.
Light travels in straight lines.
When light hits objects, it is reflected.
You see things because objects reflect light.

Shadows

Light can pass through some things.
It can not pass through other things.
An object that blocks light makes a shadow.

Heat

Heat is energy that makes things warmer.
Fuels give off heat as they burn.
Oil and wood are fuels. Natural gas is a fuel.
People use fuels to keep warm and to cook.

Friction

When objects rub together, they get warm.
The heat is caused by friction.
Friction slows down the objects.
Friction also causes the objects to get warm.

Heat and Electricity

nuclear energy station

Heat is used to make electricity.
First, heat changes water into steam.
The steam turns machines.
The machines make electricity.

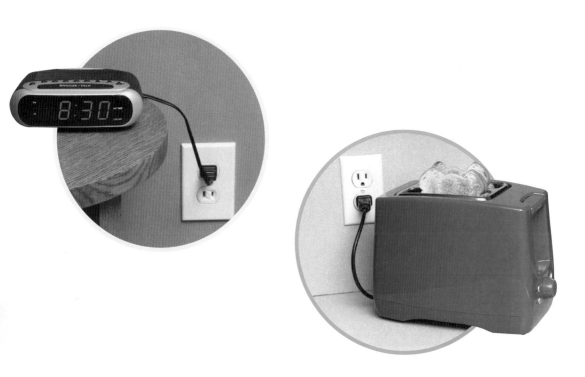

Power lines carry electricity to buildings.
Electric wires go to outlets in walls.
Some things change electricity to heat.
Other things may change it to light or sound.

Heat Travels

Heat moves from warmer to cooler things.
Heat travels easily through things like metal.
It does not travel easily through plastic.
It does not move easily through oven mitts.

Measuring Heat

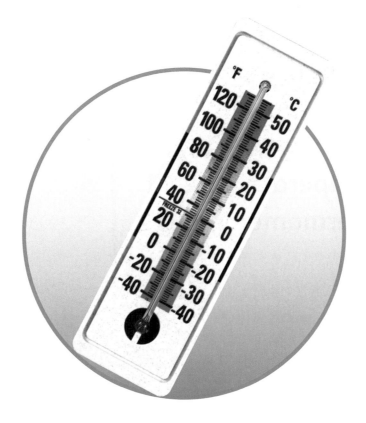

Temperature tells how hot or cold a thing is.
A thermometer measures temperature.
It may measure in degrees Fahrenheit.
Or, it may measure in degrees Celsius.

Vocabulary

electricity, p. 8

energy, p. 2

friction, p. 7

heat, p. 2

light, p. 2

reflect, p. 4

solar energy, p. 3

temperature, p. 11

thermometer, p. 11